NATIONAL GEOGRAPHIC KiDS

美国国家地理
双语阅读

Deadly Predators

危险的捕食者

懿海文化　编著

马鸣　译

第三级

外语教学与研究出版社
FOREIGN LANGUAGE TEACHING AND RESEARCH PRESS
北京 BEIJING

京权图字：01-2021-5130

图书在版编目 (CIP) 数据

危险的捕食者：英文、汉文／懿海文化编著；马鸣译 . —— 北京：外语教学与研究
出版社，2021.11（2023.8 重印）
（美国国家地理双语阅读 . 第三级）
书名原文：Deadly Predators
ISBN 978−7−5213−3147−9

Ⅰ . ①危… Ⅱ . ①懿… ②马… Ⅲ . ①野生动物－犬科－少儿读物－英、汉 ②野生动
物－猫科－少儿读物－英、汉 Ⅳ . ①Q959.838−49

中国版本图书馆 CIP 数据核字 (2021) 第 228168 号

出 版 人　王　芳
策划编辑　许海峰　刘秀玲　姚　璐
责任编辑　姚　璐
责任校对　华　蕾
装帧设计　许　岚
出版发行　外语教学与研究出版社
社　　址　北京市西三环北路 19 号（100089）
网　　址　https://www.fltrp.com
印　　刷　天津海顺印业包装有限公司
开　　本　650×980　1/16
印　　张　37.5
版　　次　2022 年 3 月第 1 版 2023 年 8 月第 4 次印刷
书　　号　ISBN 978-7-5213-3147-9
定　　价　188.00 元（全 15 册）

如有图书采购需求，图书内容或印刷装订等问题，侵权、盗版书籍等线索，请拨打以下电话或关注官方服务号：
客服电话：400 898 7008
官方服务号：微信搜索并关注公众号"外研社官方服务号"
外研社购书网址：https://fltrp.tmall.com

物料号：331470001

记载人类文明
沟通世界文化
www.fltrp.com

Table of Contents

Hungry Hunters

Grizzly bears eat fish and other small animals. They usually weigh 300 to 500 pounds. The largest bears can weigh 1,800 pounds!

Wolves chase.
Sharks attack.
Bears lunge.

All these animals are predators. They're all after the same thing.

Meat. They need it to live and grow.

Word Bite
PREDATOR: An animal that kills and eats other animals

Dogs in the Wild

Wolves are members of the dog family. So are foxes, coyotes, and African wild dogs. All of these dogs are powerful predators.

Big ears help wild dogs hear their prey. Sensitive noses help the dogs sniff out food.

Wild dogs also have large teeth and strong jaws. They can kill prey larger than themselves.

Word Bite

PREY: An animal that is killed and eaten by another animal

A coyote has many teeth. Some can tear flesh. Others can crush bone.

African wild dogs hunt in packs.

An arctic fox stalks its prey.

Cats in the Wild

Cats in the wild are all expert hunters.

The rusty-spotted cat is the smallest. It weighs less than a half-gallon jug of milk.

The rusty-spotted cat lives in India. At night, it hunts for birds, mice, lizards, and frogs.

Q What happened when the tiger swallowed a ball of yarn?

A It had mittens.

What is the biggest cat in the world? The Siberian tiger. It can weigh more than three large men.

Siberian tigers can eat up to 80 pounds of meat at one time.

Female lions hunt in a group. They work together to get a meal.

Most cats live alone. But lions live in a family group called a pride.

The males protect the group. The females do all the hunting.

Word Bite
PRIDE: A family group of lions

A cougar has long, strong legs. But it can't run fast or far. It sneaks up on prey. Then it jumps on the animal and bites its neck.

Cougars have sharp teeth. They chop prey into bite-size bits.

Polar Power

The polar bears use their sharp teeth and huge paws to catch prey.

They hunt alone.

A polar bear has one of the world's best noses. It comes in handy when sniffing out dinner.

Q What do you call a polar bear with earmuffs?

A Whatever you want. It can't hear you!

The polar bears mostly eat seals snatched from holes in the ice. But sometimes they eat small animals, too.

Mini-monsters

These predators may be small, but they sure are deadly!

Giant Water Bug
A giant water bug hunts fish and frogs, snakes and snails. It jabs prey with its mouthparts. Then it sucks out the animal's insides!

Wind Scorpion

This little critter has huge jaws and runs like the wind. It chases down termites, beetles, and even lizards. This one caught a cricket.

Short-tailed Shrew

This shrew's spit is full of venom. The shrew uses the spit to paralyze its prey.

Army Ants

A group of army ants looks like a moving carpet. The group can be as wide as a street. It can be as long as a football field. The ants kill and eat everything in their path.

Word Bites

VENOM: A liquid some animals make that is used to kill or paralyze other animals

PARALYZE: To make unable to move

Ocean Hunters

The great white shark can go weeks without eating. But when it gets hungry. . . look out!

This fierce fish has thousands of teeth. It uses them to grab fish, seals, sea lions, and dolphins.

A great white shark attacks from below. It may push up with so much power that it rises out of the water.

Orcas attack a gray whale.

Orcas have two ways of hunting. They can surround a large animal and attack as a group. Then they share the meal.

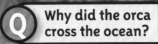

Q Why did the orca cross the ocean?

A To get to the other tide.

Orcas are called killer whales. Why? Because they eat all kinds of prey—birds, sea turtles, seals, and even sharks.

Orcas can also push a lot of fish into a tight ball. Then the orcas take turns eating the prey. This is called a bait ball.

Super-predators

Biggest

A blue whale is longer than two school buses. But it eats shrimp-like critters no bigger than your pinky!

Fastest on Land

A cheetah can run as fast as a car on the highway. It's faster than any prey.

Fastest in the Air

How does a peregrine falcon catch prey? By diving through the air three times faster than a car on the highway.

Strangest

A fossa looks like a cross between a squirrel and a kitten. It hunts lemurs, birds, crabs, snakes, and more.

Longest Jumper

A cougar has powerful back legs. When it jumps from a high place, this big cat can jump the length of a pickup truck.

Most Beautiful

A sea anemone looks like a flower. But it has tentacles with venom. They can kill fish or shrimp in seconds.

Word Bite

TENTACLE: An arm-like part of an animal used to feel things or catch food

Snakes, Lizards, and Crocodiles

A black timber rattlesnake swallows a mouse.

Snakes kill their prey in different ways. Some snakes have venom in their fangs. The venom paralyzes the animals they catch. Then the snakes swallow the prey whole.

Other snakes don't use venom. The anaconda grabs an animal with its teeth. Then it curls around the prey and squeezes it to death.

Many lizards are small enough to sit in your hand. But the Komodo dragon is as big as a surfboard.

The giant lizard has 60 sharp, jagged teeth.

The Komodo dragon is the biggest lizard alive today. It eats birds, bird eggs, monkeys, wild boar, goats, deer, and more.

Q What do you get when you cross a kangaroo and a Komodo dragon?

A A leaping lizard.

Its spit is even more deadly. It's full of germs that can kill prey in one bite.

How does a hungry crocodile catch its prey? It lies in the water and waits. The patient hunter can wait for hours.

When prey passes by, the croc grabs it, pulls it underwater, and waits for it to drown.

What does a crocodile eat? Almost anything—fish, birds, snakes, deer, wildebeest (below), and more.

Perfect Predators

Crocs, cheetahs, wolves, sharks, and snakes are some of the most awesome predators on the planet.

But they don't hunt for fun.
They hunt to feed themselves
and their families.

Cheetahs are the fastest animals on land. They eat animals such as gazelles, impalas, hares, and young wildebeest.

Stump Your Parents

Can your parents answer these questions about predators? You might know more than they do!

Answers are at the bottom of page 31.

1

How much meat can some tigers eat at one time?

A. 10 pounds
B. 20 pounds
C. 40 pounds
D. 80 pounds

2

_____ are members of the dog family.

A. Wolves
B. Foxes
C. Coyotes
D. All of the above

3

Which predator sucks out the insides of its prey?

A. A rusty-spotted cat
B. A giant water bug
C. A peregrine falcon
D. A crocodile

4

Orcas are a kind of _____.

A. Fish
B. Whale
C. Insect
D. Reptile

5

Which of these animals does not hunt in groups?

A. Polar bear
B. Orca
C. Army ant
D. Lion

6

Which predator has the most teeth?

A. Gray wolf
B. Giant water bug
C. Great white shark
D. Komodo dragon

Glossary

PARALYZE: To make unable to move

PREDATOR: An animal that kills and eats other animals

PREY: An animal that is killed and eaten by another animal

PRIDE: A family group of lions

TENTACLE: An arm-like part of an animal used to feel things or catch food

VENOM: A liquid some animals make that is used to kill or paralyze other animals

▶ 第 4—5 页

饥饿的猎手

大灰熊吃鱼和别的小动物。它们一般重达300—500磅（约136.08—226.8千克）。最大的熊体重可达1,800磅（约816.48千克）！

狼追击。
鲨鱼袭击。
熊猛地向前扑。
所有这些动物都是捕食者。它们都在寻找同样的东西。
肉。它们需要肉来生存、生长。

小词典

捕食者：杀死并吃掉其他动物的动物

▶ 第 6—7 页

野生犬科动物

　　狼是犬科动物家族的成员。狐狸、郊狼、非洲野犬也是犬科动物。这些犬科动物都是强大的捕食者。

　　大大的耳朵有助于野生犬科动物听到猎物的声音。灵敏的鼻子有助于犬科动物嗅到食物。

　　野生犬科动物还有很大的牙齿和强壮的颌。它们能杀死体形比自己大的猎物。

小词典

猎物：被另一只动物杀死并吃掉的动物

郊狼有很多牙齿。一些牙齿可以撕肉，另一些牙齿可以嚼碎骨头。

非洲野犬成群捕猎。

一只北极狐悄悄靠近它的猎物。

▶ 第8—9页

野生猫科动物

野生猫科动物都是捕猎专家。

锈斑豹猫是最小的野生猫科动物。它比半加仑（约1.89升）一罐的牛奶还要轻。

世界上最大的猫科动物是什么呢？西伯利亚虎。它比三个壮汉还要重。

锈斑豹猫生活在印度。在晚上，它猎捕鸟类、老鼠、蜥蜴和青蛙。

西伯利亚虎一次可以吃掉80磅（约36.29千克）肉。

▶ 第10—11页

母狮成群捕猎。它们齐心协力以获得食物。

小词典

狮群：狮子的大家庭

大多数猫科动物都独自生活。但狮子生活在被称为"狮群"的家庭里。

公狮负责保护狮群。母狮负责捕猎。

美洲狮的腿又长又壮。但是它跑不快，也跑不远。它悄悄靠近猎物，然后扑到猎物身上，咬住它的脖子。

美洲狮长着锋利的牙齿。它们把猎物咬成可以入口的小块。

▶ 第 12—13 页

极地力量

北极熊用锋利的牙齿和巨大的脚掌捕捉猎物。
它们独自捕猎。
北极熊通常以从冰洞里捕捉到的海豹为食。但有时它们也吃小动物。

北极熊有世界上最灵敏的鼻子之一。它们的鼻子在寻找食物时特别管用。

▶ 第 14—15 页

迷你怪兽

这些捕食者可能很小，但它们确实很危险！

田鳖

　　田鳖捕捉鱼、青蛙、蛇和蜗牛。它将口器刺入猎物的身体。然后它把猎物的内脏全都吸光！

风蝎

　　这种小动物有巨大的颌，跑起来像风一样快。它追捕白蚁、甲虫，甚至蜥蜴。这只风蝎捉到了一只蟋蟀。

短尾鼩鼱

　　这种鼩鼱的唾液里满是毒液。鼩鼱用唾液来麻痹猎物。

行军蚁

　　一群行军蚁看起来就像一块移动的毯子。蚁群可以像街道那么宽。它可以像足球场那么长。这种蚂蚁杀死并吃掉路上所有的东西。

小词典

毒液：一些动物分泌出来的、用来杀死或麻痹其他动物的液体

麻痹：使无法活动

▶ 第 16—17 页

海洋猎手

大白鲨可以数周不进食。但当它饿了的时候……小心！

这种凶残的鱼有数千颗牙齿。它用牙齿捕捉鱼、海豹、海狮和海豚。

> 大白鲨从下向上进攻。它能用很大的力量向上推起，以至于冲出水面。

▶ 第 18—19 页

虎鲸有两种捕猎方式。它们会集体围攻一只大型动物，然后分享美食。

虎鲸也可以把很多鱼赶到一起，让它们挤成一个球。然后虎鲸轮流食用这些猎物。这个球也叫"诱饵球"。

> 虎鲸攻击一只灰鲸。

> 虎鲸也叫"杀手鲸"。为什么呢？因为它们吃各种各样的猎物——海鸟、海龟、海豹，甚至是鲨鱼。

▶ 第 20—21 页

超级捕食者

最大的

蓝鲸比两辆校车还要长。但它吃像虾一样的、没你的小指大的动物。

陆地上最快的

猎豹可以跑得像高速公路上的汽车一样快。它比任何猎物都跑得快。

空中最快的

游隼怎样捕捉猎物呢？它以比在高速公路上行驶的汽车快三倍的速度在空中俯冲。

最奇怪的

马岛狸看起来像是松鼠和小猫的合体。它猎捕狐猴、鸟、蟹、蛇以及其他动物。

跳得最远的

美洲狮有着强健的后腿。从高处往下跳时，这种大型猫科动物可以跃过一辆小卡车。

最美的

海葵看起来像一朵花，但是它的触手上有毒液。它们能在几秒之内杀死鱼或虾。

小词典

触手：动物身上像手臂一样、用于感知物体或捕获食物的部位

▶ 第 22—23 页

蛇、蜥蜴和鳄鱼

蛇用不同的方式杀死猎物。一些蛇的尖牙里有毒液。这种毒液能麻痹它们抓到的动物。然后蛇就把整个猎物吞下去。

另一些蛇不用毒液。水蚺用牙齿抓住动物。然后它用身体缠住猎物，把它挤压死。

一条黑色森林响尾蛇在吞食一只老鼠。

▶ 第 24—25 页

许多蜥蜴很小，可以坐在你的手上。但科莫多巨蜥却像冲浪板那样大。

这种巨蜥有 60 颗锋利的、锯齿状的牙齿。

它的唾液更危险，里面满是病菌，可以一口将猎物置于死地。

科莫多巨蜥是现存体形最大的蜥蜴。它吃鸟、鸟蛋、猴子、野猪、山羊、鹿以及其他动物。

▶ 第 26—27 页

鳄鱼吃什么呢？几乎什么都吃——鱼、鸟、蛇、鹿、角马（如下图所示）以及其他动物。

饥饿的鳄鱼如何捕捉猎物呢？它趴在水中等待。这个耐心的猎手可以等上好几个小时。

当猎物经过时，鳄鱼抓住它，把它拖到水下，直到它溺死。

▶ 第 28—29 页

完美的捕食者

鳄鱼、猎豹、狼、鲨鱼、蛇是地球上最可怕的捕食者中的一部分。但它们不是为了好玩儿而捕猎。它们捕猎是为了喂饱自己和它们的家人。

猎豹是陆地上速度最快的动物。它们吃瞪羚、高角羚、野兔和小角马等动物。

▶ 第 30—31 页

挑战爸爸妈妈

你的爸爸妈妈能回答这些关于捕食者的问题吗？你可能比他们知道的还多呢！答案在第 31 页下方。

1

有些老虎一次可以吃多少肉？
A. 10 磅（约 4.54 千克）　　B. 20 磅（约 9.07 千克）
C. 40 磅（约 18.14 千克）　　D. 80 磅（约 36.29 千克）

 **2**

_____ 是犬科动物家族的成员。
A. 狼　　　　B. 狐狸
C. 郊狼　　　D. 以上都是

 **3**

哪种捕食者吸食猎物的内脏？
A. 锈斑豹猫　　B. 田鳖
C. 游隼　　　　D. 鳄鱼

4

虎鲸是一种 _____。
A. 鱼　　　　B. 鲸
C. 昆虫　　　D. 爬行动物

 5

哪种动物不成群捕猎？
A. 北极熊　　B. 虎鲸
C. 行军蚁　　D. 狮子

6

哪种捕食者的牙齿最多？
A. 灰狼　　　B. 田鳖
C. 大白鲨　　D. 科莫多巨蜥

词汇表

麻痹：使无法活动

捕食者：杀死并吃掉其他动物的动物

猎物：被另一只动物杀死并吃掉的动物

狮群：狮子的大家庭

触手：动物身上像手臂一样、用于感知物体或捕获食物的部位

毒液：一些动物分泌出来的、用来杀死或麻痹其他动物的液体